Application of Ergonomics in Politics

Politics and ergonomics share a commonality as they are designed for people. Despite this connection, the application of ergonomic principles to political success remains largely unexplored in academic literature. This book presents ergonomics in an accessible format specifically tailored for those involved in politics, regardless of their educational or professional background.

Written in an engaging storytelling format, this guide covers six essential areas: the macro picture of politics and elections, the ergonomic approach to politics and power, ruling and opposition dynamics, connecting with people, election ergonomics and practical exercises for application. Each chapter begins with an overview section and concludes with key points for easy digestion and future revision, followed by practice sessions that allow readers to test their understanding. This structured approach ensures that readers can gain practical skills they can immediately apply to enhance their political effectiveness.

Application of Ergonomics in Politics: A Layperson's Approach is designed for those in the field of politics including politicians at all levels, political campaigners, public speakers, election organizers, political science students and anyone interested in understanding how ergonomic principles can improve political processes. It will also appeal to practitioners of ergonomics and human factors in understanding where ergonomics and politics intertwine.

VOTE

Application of Ergonomics in Politics

A Layperson's Approach

Prabir Mukhopadhyay, MSc., PhD.

CRC Press
Taylor & Francis Group
Boca Raton London New York

CRC Press is an imprint of the
Taylor & Francis Group, an **informa** business

Designed cover image: Prabir Mukhopadhyay

First edition published 2026
by CRC Press
2385 NW Executive Center Drive, Suite 320, Boca Raton FL 33431

and by CRC Press
4 Park Square, Milton Park, Abingdon, Oxon, OX14 4RN

CRC Press is an imprint of Taylor & Francis Group, LLC

ISBN: 9781041251187 (hbk)
ISBN: 9781041258650 (pbk)
ISBN: 9781003748229 (ebk)

DOI: 10.1201/9781003748229

Typeset in Times
by codeMantra

To the Lotus Feet of Shri Ramakrishna Paramhansha Dev, Pronam To You...

Late Dhirendranath Mukhopadhyay (My Father), miss you!!

&

Mrs Meena Mukhopadhyay (Mother), Dada, Mamoni, Duggi and Leto…

Contents

List of figures *ix*
About the author *xi*
Preface *xiii*
Acknowledgements *xv*

1 The macro picture of politics and election **1**
Overview 1
1.1 The systems approach in politics and election 1
1.2 The weak link in the system from ergonomics perspective 3
1.3 Making the political weak link robust through ergonomics 5
1.4 Parts of the political system: an ergonomic perspective 6
1.5 The politician: a human and the human factors 7
1.6 What ergonomics has to offer and position in the political ecosystem 8
1.7 Key points 10
1.8 Practise session 10

2 Politics and power: the ergonomic approach **12**
Overview 12
2.1 The human characteristics related to power play 12
2.2 Living and working with people at grassroots level 14
2.3 The cognitive ergonomic approach of hierarchy in politics 14
2.4 Gaining, regaining and retaining political power through ergonomic approach 15
2.5 The ergonomics of political violence and its prevention and control 16
2.6 The humane approach in influencing people in favour of an ideology 17
2.7 Key points 18
2.8 Practise session 18

3 The ruling and the opposition and ergonomic angle **20**
Overview 20
3.1 The divide between ruler and opposition 20

3.2 Peaceful protest and the cognitive ergonomic issues 23
3.3 Perception creation by the two groups among the general public 24
3.4 How to be relevant as opposition through ergonomics 25
3.5 How to ensure that ruler sustains in power through ergonomic intervention along product, space and communication 26
3.6 The personal grooming of politicians and image building 27
3.7 Key points 28
3.8 Practise session 29

4 Connecting with people **31**
Overview 31
4.1 The needs and wants of people 31
4.2 Product, space, communication and service: avenues for ergonomic intervention 33
4.3 Communication through visual ergonomics 34
4.4 The communication needs among rich and poor, rural and urban areas 35
4.5 How to deal with a mixed population 36
4.6 Different modalities of communication: ear, eyes, nose, tongue, touch 36
4.7 Key points 37
4.8 Practise session 38

5 Ergonomics of election **39**
Overview 39
5.1 Election campaign and visual ergonomics 39
5.2 Touching the electorate through visual ergonomics 41
5.3 The election process and the ergonomic approach 43
5.4 Cordial relationship in election: the ergonomic angle 45
5.5 Ergonomics in preventing election corruption 46
5.6 Using ergonomics to make an impact during campaign 47
5.7 Key points 48
5.8 Practise session 48

6 Exercises on application of ergonomics in politics **50**

Bibliography *57*
Index *61*

Figures

1.1	Ergonomic issues in the political system of the country Eldorado	3
1.2	Importance of feedback in existence of political party	4
1.3	Ergonomic aspects of the political system	6
2.1	Power pyramid	15
3.1	The ruler and the opposition	21
3.2	Good mental image of a political party	22
3.3	People's protest is like pressure building up in pressure cooker	24
3.4	Product space and communication in politics	27
3.5	Personal grooming in image building in politics	28
4.1	Nitai working at grassroots level to get important information before its lost	32
4.2	Citizens' misunderstanding at common facilities like toilet in public places	33
4.3	Political messages and its accuracy	35
4.4	Repeating political tag line through audio and video modes to reinforce the same in the citizens	37
5.1	Visual ergonomics in election campaigns	40
5.2	Interacting with the electorate through ergonomics	42
5.3	Visual ergonomics in touching hearts	42
5.4	Human body dimensions for designing voting booths	44
5.5	Interactive seating arrangements facilitating harmony during elections	46
5.6	Election campaign for encouraging citizens to vote	47

About the author

Prabir Mukhopadhyay (Associate Professor, Design Discipline, Indian Institute of Information Technology Design and Manufacturing Jabalpur, India.)

Prabir Mukhopadhyay holds a PhD in Industrial Ergonomics from the University of Limerick, Ireland. Prabir started working with noted ergonomist Prof R. N. Sen, at the Calcutta University both for his Master's Thesis and later on a project sponsored by the Ministry of Environment and Forests, Government of India. It was during this time Prabir developed a keen interest in the subject and wanted to build his career in ergonomics. He joined the National Institute of Design, Ahmedabad, India, as an Ergonomist for one of the projects for the Indian Railways. He then joined the same institute as a faculty in ergonomics. During his tenure at Ahmedabad, he worked on many consultancy projects related to ergonomics. Some of his clients there included the Indian Railways, Self Employed Women's Association and the United Nation Industrial Development Organization.

After working there for two years Prabir left for the University of Limerick, Ireland, on a European Union Funded Project under the supervision of Prof T. J. Gallwey. He completed his PhD in Industrial Ergonomics from the same university and decided to return to India to apply his acquired knowledge. He joined the National Institute of Design, Post Graduate Campus at Gandhinagar, India, as a faculty in ergonomics. There he headed the Software and User Interface Design discipline. After working there for around five years, Prabir joined his present institute as an Assistant Professor in the Design Discipline. He was later promoted to an Associate Professor. He was the Discipline Head for nine years. He has authored seven single-authored books till date. His favourite pastime is writing, travelling and cooking.

Preface

Politics is a very important area and is one of the pillars of any democracy. Every country needs good political leaders and good governance, and this in the long run ensures that peace prevails on this earth. Till date all books on politics have been written keeping in mind those with a background or interest in politics. But there is a distinct dearth of books in the market which connects or looks at politics through the lens of ergonomics or the people's perspective. This book is an attempt in that direction. It tries to explain the ergonomic issues in politics and election and gives direction as to how the common people can also participate in politics and thus contribute to nation-building. It also gives insights into the application of ergonomics in the election process for different political parties. The language of this book is in a storytelling format with hypothetical characters along with visuals, which should make reading a pleasure for the common person. Every chapter in this book starts with an overview and ends with key points and some exercises with ergonomic directions. The last chapter (Chapter 6) presents a series of problems on application of ergonomics in politics along with their ergonomic directions for addressing them. This book should appeal to the common person as well as those in active politics.

I hope readers enjoy this book, and any criticism and suggestions are most welcome.

Prabir Mukhopadhyay, MSc., PhD.

Acknowledgement

All the diagrams in this book have been drawn by two of my Bachelor of Design Students, namely, Kritika Kakkar and Akhil Negi. Thank you so much for breathing life into the text so beautifully!!

1 The macro picture of politics and election

OVERVIEW

The chapter introduces the readers to the macro picture of the application of ergonomics in the political scenario of any place. The position of ergonomics as a discipline in the political system is discussed in this chapter with specific emphasis on how to make the system robust from an ergonomic perspective. The chapter highlights the human side of politics and discusses ways and means of making the entire political ecosystem, robust, people centric and corruption free.

1.1 THE SYSTEMS APPROACH IN POLITICS AND ELECTION

When two or more elements work together or interact with one another for a common goal or a common purpose it's called a system. The entire fabric of politics and political system is also a systems concept, as there are different people, a variety of contexts and election, and it has its further subdivisions which all work in a synchronised fashion for giving the desired benefit.

DOI: 10.1201/9781003748229-1

The story is of a country Eldorado. It's a small country where the political party "kichir michir" is in power for the last two terms. Fortunately the people are extremely happy and they want the current political party to stay in power. The reasons are many. The citizens believe that they are being well looked after. Their welfare and well-being are taken care of. At the time of election the opponent political party "Dhyat Teri" did a thorough research to find out some loopholes so that the same could be highlighted in the election campaign. Unfortunately they failed. Detailed analysis from an ergonomic perspective revealed that the current political party has designed the administrative system so well that people had access to all officers. There were neither bottleneck for the public when they wanted to get a service nor the public were made to wait for long. Every query of the public was listened to, and they were explained clearly how fast their problem would be resolved.

The interesting part was that the public was allowed to participate in the decision making process of the government and could give their views. The communication between the political leaders and the citizens was very frequent and clear, and the citizen's voice was respected. The political leaders were also easily accessible to the citizens, and they got them whenever they needed. In other words, citizens enjoyed the rule of the political party because of various reasons which were in essence "in favour" of the citizens at large.

If we look at the above story of Eldorado, we will find that there are lots of "favourable" issues that played a key role in the parts popularity, and hence they were voted to power (Figure 1.1). First issue was absolute transparency. It was clear to citizens where to reach for any problem and the status of the complaint after it was lodged. Second was there were no bottlenecks or obstructions for the citizens if they wanted to interact with any official or leaders of their locality. The voting mechanism was also strong, and at every election the ruling party never tried to manipulate or threaten the people. They tried to win the heart of the citizens by communicating to them what they have done, what they could not do and what are their plans if they came to power.

So the key to their success was efficient communication with the citizens and gifting to them a "transparent" system where everything was crystal clear with no ambiguity of any nature. Citizens are humans, and they always expect a "feedback" from any system which was very much present. Any system should be transparent so that users know whether things are going wrong and where is the source of the error. This helps in recovering from an error or mistake, thus making the administrative and political system robust.

We need to remember that in a system for the benefit of the citizens, the citizens are the most important part. Administration and political party

FIGURE 1.1 Ergonomic issues in the political system of the country Eldorado

is there to augment the system. The interface between citizen and administration should be seamless so that the citizens get a wonderful experience and get that satisfaction with the administration and the current political system.

1.2 THE WEAK LINK IN THE SYSTEM FROM ERGONOMICS PERSPECTIVE

In any country, there is an administration comprising administrative staffs, there are different ministers belonging to the ruling political party and there are the citizens. These comprise the administrative system. Citizens are the ones for whom the administration exists. You can say citizens are the God! So if your God is not happy, then there is every possibility that the system would collapse, because in that case there would be no interaction, and hence the system would become weak at one point, thereby increasing the chances of the system collapsing. So what can make the system vulnerable to failures and collapse? It's always the citizens. Remember that every citizen is different from each other. The difference is in education, age, perception, aspiration and characteristics, so the system has to accommodate citizens with these varied characteristics. Therefore, the biggest challenge is to serve variety of citizens as administrators and political leaders.

In a country "Beguni", the citizens voted to power the political party "Petebatha". Initially the party was extremely popular and people were happy. Gradually with passage of time the leaders became busy with themselves and stopped interacting with people. People had no clue as to how and what developmental work were happening which would address their day-to-day problems. The result was that the people became agitated and the party lost miserably in the next election. From this incident the major learnings were that any system should be transparent for the public to see what exactly is going on. It's not necessary that all problems would be solved, but sincere efforts being made for solving the problems should be conveyed to the people. This feedback is most important for any political system to survive. The second learning was that communicating to people in their language is very important. This communication by politicians should be more focussed to listening to people and less on speaking. Once the route of listening is opened up, people feel the satisfaction that their grievances have been properly communicated, and hence there is a sense of satisfaction among people. Building this image (called mental model) based upon interacting with politicians by people is very important for any political party to be successful. Remember every human is good; they become aggressive when all the routes of communication with the political party are closed (Figure 1.2).

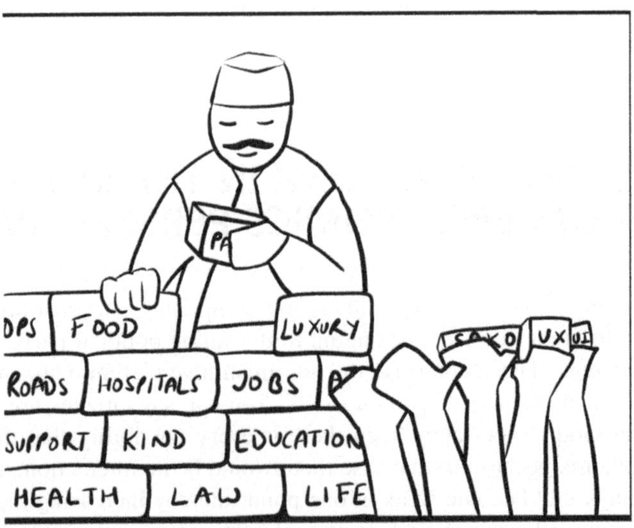

FIGURE 1.2 Importance of feedback in existence of political party

1.3 MAKING THE POLITICAL WEAK LINK ROBUST THROUGH ERGONOMICS

It's easy to point people's mistakes but when it comes to our own we are not ready to accept the same. You have heard that humans make mistakes, but can mistakes (sophisticated name is error!) be eliminated? In the previous paragraph we talked only about citizens with varied characteristics. What about administrators, leaders and ministers with similar varied characteristics? So you can understand that in an administration there is so much of variation because of different types of people on either side: citizens and administrator. Ergonomics comes to your rescue at this juncture.

Training of the administrators and the politicians in how to deal with people is beneficial. If all administrators and politicians receive the same training, then the variation among them could be reduced to some extent. For example, if the training module suggests that one should listen to the citizens even if they are arrogant, then it's expected that all administrators would deal with citizens not in their own way but similar way. But on the side of the citizens how the variation could be reduced? First, remember to deal with each citizen separately. Make sure the space or the context where the communication takes place is comfortable and has the basic amenities and facilities like drinking water, toilets and adequate seating for all. The entry and exit to the space should be guided by proper information system so that people don't get irritated in finding out the contact person. Amenities ensure that the basic facilities are provided to the citizens and they are at comfort. Now automatically once comfortable your anger would subside!!

Nimai is an 80-year-old retired school teacher. It has been two months that he has not been receiving his pension. He contacted the administration over the telephone many times but could not get a reply as to what exactly to be done. So he decides to go to the administrative office. He keeps asking people in the office as to who should he exactly meet for handling his unique problem. Nimai receives no satisfactory answer, and after spending a whole day he is frustrated and comes back. His neighbours advise him to reach to the local leader of the ruling political party. Nimai meets the leader Ms Mami. Mami was courteous and listened to Nimai. She assured him that she would look into the matter. Accordingly, she gives Nimai the contact number of a person in the administration. Nimai again visits the office and this time meets the contact person directly who finally solves the problem. Nimai is terribly frustrated by the way things worked for him! Weak link in the political system develops only when the citizens suffer due to lack of information and transparency in the system. People expect their problems to be sorted at one go. If the same problem is sorted out in the second attempt (in this case of Nimai going to the leader and again to the same office), it's irritating and etches a different mental model of the situation in people's mind. We all know that humans remember pain points more vividly than the pleasures in life. People always remember the first experience, and even if their problem is sorted out, the first experience is always seen as a great potential loss of time, energy and resources. This leaves a lasting impression in citizen's mind which is later reflected in the vote bank.

1.4 PARTS OF THE POLITICAL SYSTEM: AN ERGONOMIC PERSPECTIVE

The political system (Figure 1.3) from an ergonomics perspective comprises many parts. Citizens are the most important because it's they who bring and oust a political part from power that they deem fit. So it's important that the ethos, moral and objectives of the political party are clearly articulated to citizens in a language that they understand. The leaders are role models. Citizens prefer to see themselves in them. They are another part of the political system. Any hierarchy in the system should be clear to the citizens who should be free to approach anyone in the hierarchy. The purpose of hierarchy should be to allocate functions to different leaders so that power is decentralised. In this way, excess load on any one leader is reduced and chances of political mistakes are reduced. If at all there are errors made it can be rectified at the local level of the politicians. The initial frustration of citizens is always seen and perceived as critical and painful and people remember them more than anything else. It goes to the long-term memory of citizens. So in the political system care should be taken to ensure that the initial interaction with citizens is the best one so as to leave a long-lasting impression in the minds of the citizens. Services rendered to citizens in time of need are important. For example, at the time of any pandemic if a political party comes forward and helps the citizens, it makes the bonding with the citizens robust and reduces vulnerability.

FIGURE 1.3 Ergonomic aspects of the political system

1.5 THE POLITICIAN: A HUMAN AND THE HUMAN FACTORS

Politicians are also humans. Hence, all the normal human characteristics can be expected in them. Some of these are forgetfulness, making error and expecting feedback and at times reconfirmation about a particular act from the citizens. Citizens on the other hand also have the same characteristics.

Politicians should have an excellent opportunity mapping capability. Only problem-solving ability would not work, because problems have a "problem" in itself. Sounds weird? Let me explain. Your problem might not always be my problem! I am feeling hot in the room but it's possible that you are feeling cold. So if you go by problem-solving, then whose problem are you going to solve? The person feeling cold or the person feeling hot? So a better way is to look for opportunity mapping as a politician wherein you look for opportunity where you can intervene and benefit people.

You as a politician should have the ability to visualise what can happen, go wrong and interference required. This power of visualisation in advance makes a good politician and can be inculcated through extensive reading, talking to people at grassroots level and through self-introspection.

Politicians have to keep themselves physically and mentally fit to withstand pressures of long and odd working hours. They need to handle tremendous mental pressure. This is where they need to perform physical exercises and inculcate certain hobbies like music and gardening which would keep themselves physically and mentally fit and act as great stress busters.

One has to be a good communicator with the citizens. Speak short sentences; slowly and if possible communicate to citizens through real-life stories. Citizens will listen to you. Do not always highlight how great you are, rather project opportunity areas for people where you would like to intervene. People should visualise that what you are saying is their needs and wants as well. As politicians you should always be with people at the time of their need. You should be able to share their pain and agony and try putting yourself in their feet. This would project you as a true politician.

Politicians should project their image as role model which people would aspire to become. Thus, be honest and true to yourself. Your image in real life should create an image in the mind of the people and should symbolise epitome of honesty and sincerity. Here your attire which ought to be simple and sober plays a significant role. A politician in a red trouser is perceived as a joker. The same who is elegantly dressed up in formal wear is respected and touches people's heart.

There was a political leader in the country of "Bhalobasha". He was the people's leader because of his image. His attire was unique in the sense that both the shirt and the trousers are white in colour. He wore the cheapest watch and always travelled in an old bicycle. He stayed in a thatched hut and ate the simplest of food. This image of the leader named "Vipul" was iconic which was etched in the people's mind as the picture of a true leader, the people's leader. He was an epitome of honesty, simplicity and truthfulness. His life was transparent in the sense that the common people had access to his house, food, lifestyle and everything. Thus, between Vipul and the people, there was seamless communication and perfect transparency as the people could "see" Vipul and his way of life. This creates an image, an image of a true leader. The political party he represents then is also tagged with his image. It is like two people are very good friends only when they share similar characteristic features. This makes an impact on election and the electorate.

Hooliganism in politics is a means to terrorise people and try winning their hearts which never happens. Humans are sensitive, ready to forgive but they expect handholding at the time of crisis which they remember forever and thus develop alliance for a particular political party. Politicians should remember to be with people and not dole out a few penny to win their hearts. The very act of the politicians is keenly observed by common people which creates a mental model (mental image) in their minds, and this is the key to success of any politician and good political party.

1.6 WHAT ERGONOMICS HAS TO OFFER AND POSITION IN THE POLITICAL ECOSYSTEM

The dimension of different spaces, like the administrative buildings and political party offices, should factor in human body dimensions and movement so as to ensure seamless movement of people and their comfort. Illumination and ventilation are the keys in facilitating effective communication between the citizens and politicians. If you want to make an impact, then visuals of political ethos in the space play a very important role as well.

People's work should not suffer. There should be seamless workflow without any bottleneck, and this is where time plays an important role. Citizens should be given adequate feedback as to whether their concerns are being addressed and if yes to what extent.

Political ethos, will and decisions should be transparent and be visible to all, especially the citizens. It should seamlessly flow from the political hierarchy to the different strata of the administration with the general public made aware of the same. Remember that people are the key and the survival of the politicians depends on the people.

Citizens need to clearly know the roles and responsibilities of different leaders and administrators and when to approach whom for a particular purpose. If necessary have hierarchical charts at strategic locations in government buildings where the citizens would know who is what and where to find whom.

The political party named "Lajegobore" is contesting for the first time from the constituency of "Jatra Pala". The leadership of the party is tensed as to their results. They have taken. A new strategy that not only they would focus on the developmental work for the people, but they would also focus on making the entire political system transparent and robust for the people even if they lose the election. They tailer their campaign in a completely different way focussing on the ergonomic aspects in the election process. The first thing that they started projecting is a model of an ideal voting centre where people don't have to stand long for voting and how they would get a feedback that their vote has been cast in the favour of their desired candidate or the political party. The model also highlighted a voting centre with enough space for people to move, especially the elderly with all furniture pieces designed as per human body dimensions along with proper display of information. Supplementing this model was another model explaining how to make the voting process fair and transparent. This was depicted through the example of a big white swan swimming across a lake. The swan represented transparency, honesty and corruption free. The lake represented the election process where citizens could vote freely without fear. The third model depicted a model government office where an elderly couple goes for their pension-related issue and their problems are solved at the very first point without them to wait for hours. This display and depiction appealed to the people, and "Lajegobore" won the election with massive victory margin.

So in essence Lajegobore party was successful in projecting to the people an "image" of an organised system and society in which they would be a part. People always like to dream big and see themselves living a happy, decent and corruption-free society. People like to forget their past and look forward to a better future. They have a bad short-term memory (like the computer's random access memory) wherein they forget their political experiences very fast and look forward to a promising future. If these aspects of the people can be taken care of, any political party would be at an advantage. This bigger picture of politics and election needs to be understood by everyone. If you as a politician know where your people are, what they would like to see and where they would like to take their future to and if they are assured of a free and fair election process, the voters are and will always be with you.

The flow of any information from the party to the people should be seamless and in a manner people understand. For example, if the political party is organising a free medical checkup in remote rural areas, it should be displayed in public not only in rural areas but also in urban areas so that people get to know the good work done. These points are registered in people's brain which they later retrieve while voting and comparing other political parties with. Similarly, any deliberations within a political party including inner party elections should be made public even though it doesn't concern the public. This is to ensure easy availability of information about the political party in public domain which makes the people think that the party is communicating their functioning to them and hence people feel they are being given importance. Thus, the liking for the party increases as people are given a peek into the interior of the political parties' functioning. This preview of what's happening inside a party acts as a feedback to the public and increases the party's reliability and image in public domain.

1.7 KEY POINTS

A. Politics is complex and has many parts interaction with one another for a common goal, and this is called system.
B. To become successful, politician ergonomics suggests transparency in actions.
C. Always involve people in decision making, the more you are with people the better politician you are.
D. Efficient communication with citizens makes successful politicians.
E. Look for opportunity for the benefit of people and not for problems.

1.8 PRACTISE SESSION

A. Political party "begun pora" has lost the previous election in local civic poll. Your advice is solicited as a consultant ergonomist to help them come back in the next election.

Ergonomic approach

1. Systems-level analysis to be done to check weak link in the party structure
2. How is the interface with citizens? What manner
3. Work done for people and its projection in what format? Verbal, visuals, etc.
4. Where is the opportunity for people's benefit which would make maximum impact?

B. A political party "kochu pora" is gradually gaining popularity as the opposition. The next election is far away, around three years down the line. The party leaders are soliciting your ergonomic advice as to how to gain power and for the next government.

Ergonomic approach

1. Project some themes linked with people's benefit
2. Articulate clearly what are the agenda. Be precise…
3. Project leaders, their qualifications and especially lifestyle which are similar to the citizens
4. Don't make tall promises rather make small time-oriented promises.

Politics and power

The ergonomic approach

2

OVERVIEW

The chapter gives a brief overview of what power in politics means from an ergonomic view point. The different human characteristics linked to people and power relationship with respect to politics are discussed here, including how the power game is played in politics, the hierarchy of power, from an ergonomic perspective. The ergonomic approach in transfer and dissemination of power in politics is discussed. The chapter ends with the ergonomics of gaining, regaining and losing power and the different aspects of political violence which are connected with power.

2.1 THE HUMAN CHARACTERISTICS RELATED TO POWER PLAY

Mr Kachkola da is a popular and prominent figure in politics for the small country of "Telebhaja". He was voted to power in the last election by the people. In the ministry he holds an important portfolio as the Prime Minister of his country. Mr Kachkola da has a number of ministers handling different

DOI: 10.1201/9781003748229-2

portfolios in his ministry. He thought he was doing very well and trusted his ministerial team. So for five years he never travelled to different parts of the country and always was there at the headquarters, mainly for interaction with different ministers and setting forth different policy decisions for the country. As the prime minister he was very confident that he and his party would be back to power in the country and as per the constitution of the country would again rule for another seven and half years. As such in the country everyone talked about Kachkola Da, and it seemed that people believed him to be a great leader who is ready to contribute towards development of the country and the people.

The elections were held as per schedule. The expectation of Telebhaja party was that they would get the majority and come to power. The elections went on quite well, and Telebhaja was confident of getting a majority once again. Unfortunately when the election results were declared, it was found that the Telebhaja part was defeated and could not manage to win a single seat. The verdict was in favour of a less favourite party "Do Peyaza". Everyone was shocked at this result of the election. If you look at it from an ergonomic perspective, then it would give us different insights.

There are certain human characteristics that need to be factored in wherever there are issues of communication among people, people and political party and communication between the political party manifesto and the common citizen. If citizens do not get feedback from leaders as to the progress in development of their areas, then leaders would lose respect among citizens and part members. So one of the key elements to highlight to remain in power is what I am doing as a leader with reference to what was done by others. You have to project to people how difficult the journey is to get approvals and funds for development work for people and more so at what stage of development for the people the work being done are at. Leaders become powerful in politics when they are able to communicate properly with the masses. This communication between human has to be two ways wherein the problems of the people have to be addressed. Once this is successfully done leaders become popular and powerful, as people are directly able to get a feedback of their developmental work. Power does not come by mere money and muscle force. Any power acquired by force leads to resistance from people, because this is against the population stereotype. If leaders force people to do something (e.g., give bribe for getting a work done), it's against normal mental model in task performance, and hence people would physically or mentally oppose it. If people are forced against their mental model, it's reflected in the vote box and that's what happened in the above case. Power comes to leaders because the people give them their collective powers. Thus, transparency in the system with regular feedback from the leaders is important.

2.2 LIVING AND WORKING WITH PEOPLE AT GRASSROOTS LEVEL

Humans always love coming in close contact with other humans and interacting with them. It's essential that leaders should communicate with people at grassroots level. There is a mental model (mental image) of people at the grassroots level that leaders are a different class of elite people and not aware of the hardships involved in their daily life. This mental model (you can call perception) needs to be broken by leaders by working in close contact with people at grassroots level. Remember out of sight is out of mind. People remember leaders through their deeds and through repeated exposure with them. Thus, repeated interaction with the public at grassroots level by leaders by being a part and parcel of their daily life creates a nice image of leaders in people's mind that someone has come to share with them their daily struggle. This type of people-to-people interaction leads to building of an emotional interface or bridge through which exchange of thoughts, ideas and what to do pop up. These act as guide for the leaders to focus on areas that they have to work and thus they come closer to people and become heroes to them and not just men/women with muscle power.

Nitai is a leader of the "Go hara" party. Nitai is a local leader of the village administration and is a popular leader. He stays in people's houses, dines with them and carries on his work as a leader. He has an image of a Good Samaritan among the people. In the elections Nitai always wins because his close contact with the common people through dinning and staying in their houses makes him a family member of the different families. It has become a common feature that people are used to. Because of the frequency of his visit to different homes, he has been able to create an image (mental model) among the people that Nitai is our "boy". This aspect gives him dividend during election. People always want their own people in power because they feel that they can truly represent them.

2.3 THE COGNITIVE ERGONOMIC APPROACH OF HIERARCHY IN POLITICS

Politics is a place where there is hierarchy. There is a party president followed by different dignitaries down the ladder to the party worker at the lowest level. It's essentially a triangular structure where the top leaders remain at the

FIGURE 2.1 Power pyramid

maximum distance from the leaders at grassroots level. The more the distance the more dangerous it's for the party because in that case there is minimal communication between the two. If party leader does not communicate to the workers at the lower level and vice versa, then a dissociation can arise when it comes to taking any decision. It can so happen that decisions taken at higher level can seem to be imposed on the lower level. Any failure in the party could be blamed upon the top leadership only because they are the ones who get the maximum prominence. This should be prevented if communication is facilitated between all the layers of the triangle mentioned in Figure 2.1.

2.4 GAINING, REGAINING AND RETAINING POLITICAL POWER THROUGH ERGONOMIC APPROACH

There is a saying in politics that retaining political power is the toughest of all jobs. When leaders are in power they are too close to people and people take them for granted. As variety is what humans seek, after some time people might feel bored with the leaders and expect change with a new leadership? People look for change but at the same time drastic change around them which does not fit their belief system is not warranted. People in general will develop a model of development with the existing leader's style of working, but at the same time would opt for a change. Thus, power retention or gaining for the first time depends to a great extent on what image political leaders are able to portray to the people that they would be able to visualise and relate to their lives.

The political party "Jobuthobu" in the country of "Talapoka" was aspiring to win for the fifth consecutive term, as they were overconfident about their popularity among the citizens. Unfortunately when the elections happened and the results were declared, it was found that to everyone's surprise the party "Talapoka" lost the election and hence were forced to sit in the opposition in the parliament. This was because people were expecting more from the party which was performing. When people's expectation increases, then the party on a winning spree becomes their own competitors. People's mental model projects that they are going to do something great, and when in reality such "great work" does not happen, people throw them out of power and with the same expectation bring another party to power.

2.5 THE ERGONOMICS OF POLITICAL VIOLENCE AND ITS PREVENTION AND CONTROL

People become violent when they feel threatened. This feeling can come when their life is in danger or their identity is in danger. This violence needs to be prevented but the question is how. From an ergonomic view point if it could be projected to those who orchestrate the violence that there is no threat of losing power or losing one's life. Visual ergonomics could play a very important role here. If at prominent public places (including those who take part in violence) it could be depicted that what exactly is power and how remaining, loosing and regaining power is a natural and perpetual process through which a political party only becomes matured and regains enough strength to come back to power once again. This visualization among the politicians is important. Once it's achieved any violence can be stopped to a great extent.

There are places where there are lot of violence and there are where there is none. It's only because of lack of awareness of what power exactly is that people fight among themselves. If poster, banner and movies could represent the importance of human life and property and depict to them the beauty of life and portray life and society in a holistic manner, violence is bound to reduce. When probability of violence increases, areas could be divided into smaller blocks and cordoned from one another either through barricades (in the form of thorny plants, metal barricades, trenches) which hinders movement of people and reduces the strength of violence. The very visual appeal of physical structures or its form discourages movement of people, and hence violence is precipitated.

In a small village "Sundari", there has not been a single incidence of political violence for a long time. Elections are always peaceful. The reason for the same if analysed from an ergonomic perspective is that the local political party has been able to involve the common people in all decision making, including those belonging to the oppositions. In fact, the views of the opposition are given maximum weightage. The effect is that the opposition is not in a position to oppose the government. If there are no oppositions, then the chances of violence vanish. People participate in violence only when they are not heard, or their queries addressed. People and opposition want a system which is "transparent", especially when it comes to decision making. If this vulnerable point of decision making is made transparent, violence vanishes. That's exactly what happened in this small village.

2.6 THE HUMANE APPROACH IN INFLUENCING PEOPLE IN FAVOUR OF AN IDEOLOGY

People cannot be forced to believe anything. Creating a story line is the key in people accepting an ideology. Only if people are able to relate the characters in their daily life as one of them, they will accept the political ideology of the particular political party. What makes people accept or get influenced by a particular political party? What role does ergonomics play?

It's again the work of the political party and how people can see their relevance in their day-to-day activities. The "image" or the mental model of the party, which is created by the leaders, members and the tangible outputs in the form of the work that they do, forms a collage in people's mind. If this collage is acceptable to the people and they feel or perceive that this is going to benefit them, automatically the decision making of the people in favouring a particular ideology becomes stronger. This is thus achieved in the most humane way without any force or threats whatsoever.

There was once upon a time a saint who lived on the bank of a river. She used to teach the poor children of the village free of cost, and in the evenings she used to talk to people about how humans should be kind towards one another, should not go into conflicts, love animals and so on. She was so humble and spoke so softly that people used to listen to her with rapt attention. It so happened that the entire village became an ardent fan of hers. This was possible because the saint wants to portray a picture of simplicity, honesty and service to mankind sort of image to the common people at large. This is what the people's perception of "good person at heart" is. When two perceptions (image in the brain) – the person's actual image and what they want in a person – match, the person and her ideology are accepted by the people by heart.

2.7 KEY POINTS

A. It's all about people – they are the boss.
B. Politicians should behave like ordinary people.
C. Mixing with the population is the key to political success.
D. Try wining people's heart through living examples of good deeds and good actions.

2.8 PRACTISE SESSION

A. A political party wants to ensure that they remain in power for the third consecutive term. They are a little worried about the anti-incumbency factor which might go against them. What ergonomic approach would you suggest for them?

Ergonomic directions

1. Project what has been done through a timeline; the visual representation should show the progress in public welfare.
2. Project what do they plan to do through a timeline.
3. Make no big promises to people.
4. If possible, give some simple visuals on how the party has worked better in some areas like health, education and housing compared to other parties in other areas.

B. A political party in opposition has realised that they have been harsh towards people and tried to force them towards their ideology. Now they want to regain their image among the people. Your ergonomic advice is solicited.

Ergonomic directions

1. People are very bad in short-term memory. So they forget all that have happened against them.
2. Take advantage of the above and project to people what good the party could do.
3. Project a good leader who is humble, honest and down to earth so that the person/leader projects a good image of the political party.
4. Project through a timeline and visuals project what the party intends to do if voted to power.

The ruling and the opposition and ergonomic angle

3

OVERVIEW

This chapter gives the readers an overview of the application of ergonomics in political game planning for the ruling and the opposition party at large. How ergonomics principles can be applied by both the ruling and opposition parties is discussed with some ergonomic directions for the opposition to be relevant in the political arena even after losing power. The chapter highlights certain ergonomic directions which the ruling party could adhere to in order to ensure that they sustain in power for a long period of time.

3.1 THE DIVIDE BETWEEN RULER AND OPPOSITION

Citizen's perception about a political party plays a very important role in politics. The image that the politicians portray in front of the citizens either remain in their short-term memory (just like your computer's random access memory) or may be converted into long-term memory where they are stored in an organised fashion just like books in a library (through a complex process)

DOI: 10.1201/9781003748229-3

FIGURE 3.1 The ruler and the opposition

(Figure 3.1). So what creates the difference between ruler and the opposition? It's somewhat similar to an examination. You prepare for an examination in your own way as you have no clue about the type of questions to be asked. When the examination comes you write in your own way. The results are out and you have scored very badly. Another situation is that you prepare in an organised manner with reference to previous years' questions but still you are unable to score well in the examination. Just like examination results are the indicators of your learning outcome, election results are an indicator of the work done by a political party and its visibility to the public and how well the leader is accepted by the citizens. A good student and a bad student in a class are primarily differentiated by the examination results and the marks they score in our society. This creates a divide between bad and good student.

A political party when in power should create a mental image (model) of the good things they are doing for the public and the opposition on the other has to create a mental image of what is not being done for the public. Now here comes the question of who is able to create maximum image in the minds of the people and thus influence them. One key ergonomic principle is "repetition". This means if the list of things that have been done (in brief) with selected key words are constantly presented to people either in audio visual format (best), audio format or visual format, it gets registered (Figure 3.2). People are exposed to so much of information so which one would remember and register well? You need to choose an odd time of the day when there is less flow of information to people. This might be early morning, late night and afternoons on weekends. Repeated exposure of information to people through audio, visual and audiovisual mode works to register them in their brains. Here you have to repeat till the election and throughout the year. It becomes part of their life. Make sure this is customised for different age groups. For children it could be through animation, and for adults it could be small video clips or auditory clips. Different medium should be used like television and newspapers.

There was a small town of "Luchi" with a population of only 20,000 people. The town had a very efficient local body which was governed by the political party "Golap" with a specific ideology. The party remained in power

FIGURE 3.2 Good mental image of a political party

for a long time without losing a single election for 20 years. It was a unique case to study and analyse. What did the party do to remain in power? On careful analysis you will find that they gave maximum thrust on the common people and their needs and wants. They had a year-round programme. This comprised of planting trees and arranging friendly football matches involving all the local people. Regular health and awareness camps were arranged for the citizens. In this way, regular connection with the people was established and people could interact with the leaders and narrate to them their problems. There was a free flow of information directly from people to the politician and vice versa, and no information was lost. This leads to enhanced system transparency as the people literally could see how the administration backed by the political party functioned. The political party became careful that everyone is watching them, and thus they became vigilant and the incidences of corruption were less.

3.2 PEACEFUL PROTEST AND THE COGNITIVE ERGONOMIC ISSUES

Protests by citizens and political parties are an avenue for venting out the grievances of people. It's just like a pressure cooker; as the food gets cooked, the pressure inside builds up and the excess pressure is released at regular intervals in the form of "whistle" of the pressure cooker. If the pressure is not allowed to release, it can lead to accident with the pressure cooker bursting and injuring people and property in the vicinity (Figure 3.3).

A peaceful protest was being staged by the citizens of the small country of "Telebhaja", and the main concerns of the citizens were that there was acute shortage of drinking water round the clock. It only comes at specific times of the day, an hour in the morning and an hour in the evening. They wrote to the administration and the government but received no reply to their letters.

Humans always expect feedback to any input given by them. Here their letters could be seen as an input and there should have been a feedback given to them by the administration. In the absence of any feedback which could be just one line of acknowledgement that "your application has been received and is being considered" or any feedback (here reply), the sender gets anxious, first they start wondering if it has been received, next they start wondering the "fate of the application" whether it has gone to the right place. If human keeps wondering like this and anticipates that the input has gone missing (letter or application), it adds to their anxiety and ultimately that manifest into frustration

FIGURE 3.3 People's protest is like pressure building up in pressure cooker

just like pressure build-up within the cooker. Unfortunately in the absence of any such mechanism the user gives repeated input and then the inputs become frequent and violent in the form of protest which is peaceful at the beginning but can turn violet later on.

To handle any peaceful protest is to listen to the citizens, with patience. This is the first step in the release of all pressure (agitations of citizens) without talking much. Learn to listen to people even if what they have to say is wrong, unpalatable and even loud. This act of listening to people ensures that the feedback is direct from the citizens and they get a satisfaction that they are heard from the gesture and facial expression of the administrative staffs and leaders which act as direct feedback to the citizens. Protests become violent only when all avenues of communication are severed and authorities refuse to hear the citizens.

3.3 PERCEPTION CREATION BY THE TWO GROUPS AMONG THE GENERAL PUBLIC

The game of power is all about how citizens perceive the political parties to be. This perception is the image of the political party as to its honesty, who are the people, are they good or bad, what have they done in the past and based on that what can they do in the future. Citizens look for in their leaders someone whose characteristics match theirs. Then the level of trust increases. This is the reason why leaders mix with the general people, dine with them and agree to take photographs along with them upon their request. This means of spreading among the public creates an image of the politician and the political party that they are very much their own people. In ergonomics this is called the "sharp end" of the system where citizens remain, and thus it's necessary for leaders to go to this sharp end (often referred to as the grassroots level) and interact with them. The problem is that to create a long-lasting image among citizens leaders need to repeatedly interact with citizens not just before election but throughout the year. The more the number of interactions, the better will be the image formation (perception) among the citizens. This is what the ruling as well as the opposition party needs to keep in mind in any healthy democracy.

3.4 HOW TO BE RELEVANT AS OPPOSITION THROUGH ERGONOMICS

Remaining in the opposition in politics does not mean that you are not relevant. Remember it's a wheel which is dynamic; rulers of today become the oppositions of tomorrow in any health democracy. So to become relevant in politics you have to help citizens "visualise" what they are being deprived of. If this image of deprivation is clearly etched in people's mind, they would gradually turn towards the opposition. For doing this you first need to list down the lacunae in the ruling party.

Any corruption in the society affects the citizens. For example, corruption related to health, daily food supply, education and infrastructures are the elements through which citizens interact every day. Any element or service with which citizens interact frequently aids in the development of a mental model or image of the same. Thus, any break in this interaction breaks the usual mental model and citizens get irritated. This needs to be banked upon by the opposition by pointing out the same – the lacunae of the ruling party – to the citizens.

Political leaders promise to citizens in election campaigns. If citizens visualise these promises to be relevant to their life and somehow believe them, the opposition can bank upon these promises to showcase to people how many of those promises have been kept and how many have been ignored. At this juncture the opposition can create an image of what was promised and what was not given and how the citizens were cheated. The same could be achieved much better through visuals, for example, broken roads, long queue at hospitals or admissions, so that citizens are able to visualise how their promised dreams were shattered, and thus they take a decision based on this image presented of voting against the ruling party or protesting against them.

Any interference in basic necessity for citizens causes inconvenience and anger. If this could be highlighted time and again, I could be used as a tool for campaign against the opposition to prove how useless they are and portray an image among the citizens that they should be ousted from power.

Another area where citizens are sensitive is welfare schemes during normal times and during calamities. People always remember their bad times. You don't remember your good times as much as your bad times, for example, when you met with an accident in childhood, or devastating flood in your town, and you vividly remember people who came to your rescue. People are good in long-term memory, and thus if opposition can come to the rescue of citizens during calamities, then these incidents become significant "landmarks" in their life with respect to which they are able to retrieve information around those incidences.

3.5 HOW TO ENSURE THAT RULER SUSTAINS IN POWER THROUGH ERGONOMIC INTERVENTION ALONG PRODUCT, SPACE AND COMMUNICATION

Product space and communication are the three pillars which are needed by everyone including the rulers in all spheres of the society. If a new school has to be established, one has to calculate how many students to accommodate in one class and then based upon the students' body dimensions (anthropometry), the classroom and furnitures have to be designed. So here the space and the "elements of space" – the furniture – which are essentially a product play an important role. Similarly, if a new hospital is built, the people need to know where to go and how to go. This is where a well-built communication system or proper signage and symbols that people understand plays an important role in ensuring people are able to reach where they want and that too with ease and within a short time (Figure 3.4). That means that for any public utility or developmental projects the leader does, they have to consider human body

FIGURE 3.4 Product space and communication in politics

dimensions, movements and the aspect of communication to help them navigate through the space.

3.6 THE PERSONAL GROOMING OF POLITICIANS AND IMAGE BUILDING

The attire of politicians reflects his image which is engraved in people's brain. Imagine your class teacher entering the class in a red shorts and in bare body. How do you feel? You just cannot accept this because school to you is an image of discipline and respect, and this attire does not gel with that (Figure 3.5). The same way attires of politicians are very important as they leave a lasting impression on citizens. When politicians move in public

FIGURE 3.5 Personal grooming in image building in politics

domain their every aspects of what they wear are observed by citizens. This is where the image is created. A neatly dressed politician reflects a clean person (that is the image it engraves). Whereas if the same politician wears expensive clothes and carries expensive accessories citizens wonder where and how did he get them? Corruption? And scores of doubts it creates. Thus, attires do help in image building of leaders along with their contribution to the society.

3.7 KEY POINTS

A. Citizens choose their rulers on the basis of who is better able to portray developmental work for their citizens. This depiction should be short and through audiovisual route and repeated at intervals.
B. Peaceful protests and armed protests can best be controlled by patience hearing without interference.
C. Ruler and opposition have their responsibilities in a democracy. The citizens view them or perceive them as custodians of honesty, law and order and role model. As long as these images of the rulers are intact they remain in power; when the same shifts to the opposition they become the new rulers.

D. Opposition can be relevant in politics if they portray to people the lacunae of the ruling party and how it can be improved. This should be conveyed in an easy-to-understand language.
E. Sustaining in power is possible through developmental work, with special emphasis on space product and communication.
F. Personal grooming of leaders reflects the parts image and impacts what they are in reality. They depict honesty and other good values, and it sends positive message among citizens.

3.8 PRACTISE SESSION

A. In the country "Begunbhaja" you have been assigned the task of identifying the ergonomic issues related to creating the divide between ruler and the opposition party. How would you go about handling this specific problem?

Ergonomic directions

1. Identify the macro aspects of the place and the people and how they are linked to each other.
2. Study in depth the features of the two parties that people remember and do not.
3. Identify the modality of communication by the two parties with the public.

B. Students are protesting outside the college gate, and their protest is mainly against sudden increase in the college fees. As a consultant ergonomist how would you tackle the situation?

Ergonomic directions

1. Initiate dialogue with the students and do not talk while they are telling their problems.
2. After hearing their problems list them down and repeat and cross-check with the demonstrators to ensure you heard them completely.
3. Give the demonstrators a timeline within which you would come back to them with some directions to the grievance.
4. Arrange for some refreshments for the students; it leaves an impression that you are interested to solve their issues.

C. The leader of the opposition party has approached you as a consultant ergonomist to help them gain some relevance in the political arena. What ergonomic advises would you provide them?

Ergonomic directions

1. Study the history of the party and pick up that work done by them for citizens.
2. Analyse critically where the ruling party is going wrong in terms of promise and action: use infographics and statistical data to supplement.
3. Through repeated and if possible using audiovisual route present to people the wrongdoing of the ruling party and how it should be done properly.

D. An election meeting has been conveyed in an open field. The political leaders will address the citizens. What ergonomic issues should be considered?

Ergonomic directions

1. The dimension of the stage should be calculated taking into account the number of people to accommodate. Better consider the seating dimensions of each leader which is roughly a circle with the shoulder as radius.
2. The stage height should factor the vertical and horizontal visual cone of the audience so that everyone on stage is visible.
3. There should be multiple entry and exit points in the venue for ensuring easy and faster crowd gathering and dispersion.

Connecting with people

4

OVERVIEW

This chapter gives the reader an overview of ways and means of connecting with people in the arena of politics and election. The readers are taken through the different ergonomic issues in people-to-people communication, touching upon the different communication modalities like eyes, ear, nose, touch and tongue. The ergonomic issues in communicating with people from diverse background and to convince them about the political ideology or/and make them vote in their favour through ergonomics are discussed in this chapter.

4.1 THE NEEDS AND WANTS OF PEOPLE

Identifying what citizens want and aspire is the biggest challenge for the politicians and the leaders to be in power. The entire strategy for developmental work, elections, should be geared in that direction. Humans are shy and hence getting to know these are difficult through direct and upfront questions. For getting to know about their needs and wants and then working upon that is the biggest challenge for any political party. By using ergonomic principles politicians can bank upon knowing citizens' needs and wants and address them accordingly.

The first most important step is collecting relevant data from the citizens so that it could be analysed to get some insights. This data collection could be in the form of easy-to-understand survey questionnaires and opinion pools

DOI: 10.1201/9781003748229-4

FIGURE 4.1 Nitai working at grassroots level to get important information before its lost

by the citizens, hence giving fruitful results. Another method of knowing the above is sitting with the citizens in a very cosy environment and just having a discussion with all and getting a response from them could also culminate in knowing the needs and wants. The space should be comfortable for citizens, they should be allowed to speak majority of the time and even given the liberty to counter the politicians and leaders. This approach helps in bringing out the needs and wants of the people. The furniture and temperature should be optimal to stimulate health discussion with arrangement for refreshment for the citizens which would make them feel homely and they would pour out their heart. For those citizens who have problem in moving may be asked for their opinion in online mode. Digital platforms like mobile phone or computers can be used. Ergonomic principles should be used to ensure that citizens understand what and how to answer and there is a mechanism for feedback to them at every stage. The best way to get to know the citizens from their heart is to stay and work with them. Becoming one of them and the dynamics when politicians completely mix with the citizens gives the maximum benefit of making the citizens pour out their hearts and thus giving vital clues to the politicians.

"Nitai" was working for his party (Figure 4.1), a grassroots-level party worker. After the party's severe defeat, the party wanted him to know the actual cause for defeat. Hence, Nitai packed up his bag and started travelling all over the country in public transport like a common man. He stayed in people's houses, had food with them and became their family members. Thus, the mental image that he was allowed to generate among the citizens was that of a very "dear one". Once this perception was generated people started narrating to him why they ousted his party in the election and what are those things in the party they disliked. The major complaint was the leaders were never available when citizens were in crisis. This actual picture came up only because Nitai was able to be a part of the citizens and that acted as a trigger in getting to know where the party failed to convince the people.

4.2 PRODUCT, SPACE, COMMUNICATION AND SERVICE: AVENUES FOR ERGONOMIC INTERVENTION

The next step is analysing the collected data. The data related to users' body dimensions, strength, movement and position would play an important role in designing of public places and those infrastructure where there would be movement of people along with their belongings. This would give the administrators an idea about the space requirements.

The next set of data related to how citizens think and behave can be analysed to get insights into how public services should be designed so that people understand them easily and are easy to interact. For example, the symbols used in public toilets designating male and female should be designed in a manner so that there are no confusion even among the illiterates, thus not leading to any embarrassment for a male entering the female toilet and vice versa (Figure 4.2). Similarly for devices like automatic water dispenser the user

FIGURE 4.2 Citizens' misunderstanding at common facilities like toilet in public places

should know how it is to be used or else such facilities for the public remain underutilised. Government policies should be designed in a manner that it benefits everyone. The biggest challenge here is we are all different – different in our body dimensions, ability to think and respond, knowledge level and also adaptability to a new technology. When new policies are launched the government should ensure that it benefits everyone and not one particular segment. For example, a new policy is launched wherein registration for immunization is done only through online portal. Now it can so happen that those who are not technology savvy or do not have access to digital problems might face problems. So, these people might get annoyed on the government (and hence vote against them in the election), though the objective of the government was to help the citizens.

The performance of the government is judged by the public services it provides. These services should be augmented through the application of ergonomic principles. The ergonomic principles can not only help in increasing the efficiency with which the services are delivered but at the same time would also increase public satisfaction and respect on the government. For example, in public transportation like buses a special provision for wheelchair access on the bus would be highly appreciated by not only the target users but by all as a good gesture to those who are unable to walk. Ergonomics can play a role in deciding the door dimensions and space layout inside the bus to accommodate the wheelchair.

4.3 COMMUNICATION THROUGH VISUAL ERGONOMICS

Political messages should be short and clear, using more visuals and less text. Visual ergonomic principles can guide the design of such campaigns (Figure 4.3). People visualise campaigns during day and night. For example, to facilitate better reading in low illumination it's better to have dark background with light text, keeping in mind the text length should be brief. Similarly, usage of pure red and pure green should be restricted because a portion of our population are red and green colour blind and perceive them in shades of grey. If images are used to convey a political message, make sure everyone understands them. For example, an image of a group of people with knife in hand and blood dripping from the same should depict violence. It should not be perceived as something else. This needs to be tested on the population whether they understand it as such.

FIGURE 4.3 Political messages and its accuracy

4.4 THE COMMUNICATION NEEDS AMONG RICH AND POOR, RURAL AND URBAN AREAS

When it comes to communication through digital platforms one needs to be careful how tech savvy the people are. If people do not have access to smart phones or internet, then any political campaign on these platforms doesn't reach them. It then only becomes accessible for the rich who can afford them. This homework of target users' needs should be done. Similarly, in remote rural areas communication to a large extent is dependent in physical form like banners, posters and audio due to non-availability of uninterrupted electricity and network connectivity.

If new technology for communication needs to be introduced, the users should be trained, made to gain confidence and then only they can utilise them. It takes time to develop that pattern of working with a new technology through repeated practise. The young and old alike should be trained.

Always remember that urban people are bombarded with much more information compared to rural ones. Thus, any political campaign should ensure that they are brief and repeat the campaigns again and again so that they are better registered in the citizen's brain. Rural areas where such information bombardments are relatively less do not require repeated campaigns.

4.5 HOW TO DEAL WITH A MIXED POPULATION

You need to think of a varied population as mentioned before. When using technology make sure you use the local language so that people understand. Interfaces at digital platforms should have facility for confirmation and reconfirmation so that even the person who has never used it before can use it conveniently. For campaigns make sure you use simple language and clear visuals which do not have linkage to history or historical characters which many cannot relate to and miss the message being tried to convey. For example, if the campaign is talking about global peace and is supplemented by the image of Hiroshima or Nagasaki, then ask yourself will people identify the images as those two places? If they do would they be able to relate themselves to the message of peace being portrayed? Do they know that atom bombs were dropped over these two towns?

4.6 DIFFERENT MODALITIES OF COMMUNICATION: EAR, EYES, NOSE, TONGUE, TOUCH

Information enters the body through different routes or windows, namely, eyes, ears, nose, tongue and touch. The best option when it comes to conveying information is to be multisensory. That is, convey information through visual and auditory routes and if possible through touch. How might a particular fragrance (campaign through nose) be used in the political campaign of a particular party? After repeated exposure people start relating the smell with the specific party. Maximum communication happens through posters, banners, speeches and digital platform wherein the

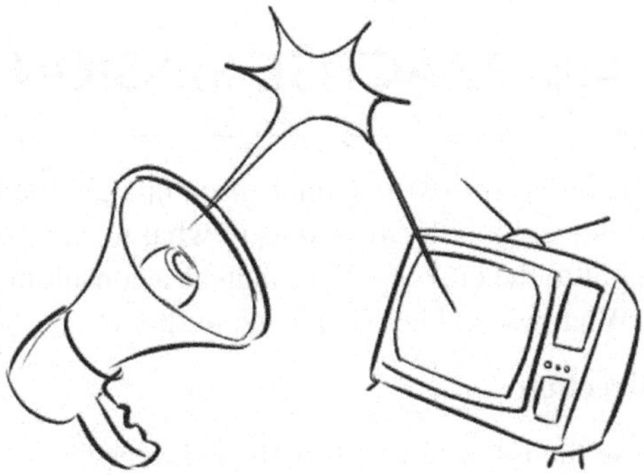

FIGURE 4.4 Repeating political tag line through audio and video modes to reinforce the same in the citizens

eye and the ears are used. The advantage is that if the eyes miss something the ears can pick them up and vice versa. Thus, the chances of the information being lost are minimised. Touch as a sense does not have much role to play in politics.

In the last election, the political party "Orebabare" did extremely well because they invested in publicity through audiovisual shows on televisions and social media platform. The repetition of the tag line of the party which was "throw out opposition" was literally transformed into long-term memory of the citizens through repeated seeing and hearing the same thing number of times throughout the day (Figure 4.4).

4.7 KEY POINTS

A. Identify people's needs and wants.
B. Think of product, space communication and services and the ergonomic intervention through these routes for the people.
C. Political messages and government messages should be brief, strategically located and understood by all.
D. Keep in mind the communications needs of urban and rural areas and the rich and poor as everyone is not exposed to same amount of information.
E. Try using maximum senses for conveying information, thereby reducing the chances of information not getting registered by the users.

4.8 PRACTISE SESSION

A. A political party wants to launch a campaign for the upcoming election. They are a bit confused as to what to campaign and issues to be brought to the citizens. Your help as a consultant ergonomist is solicited. What would be your advice to them?

Ergonomic direction

1. Analyse the issues through polls and interviews.
2. See the data for directions.
3. Design your campaign around the burning issues.

B. A political party is confused as to where and how to put the banners and flags of the party so that people are able to see them and register them. Your advice as an ergonomist is solicited.

Ergonomic direction

1. Calculate movement of people.
2. Calculate visual cone for placement of campaign.
3. You want them to read during day and night? Change background accordingly.

C. A political party has limited resources. They are looking for an ergonomic advise as to how their campaign should be designed so that they are able to address the rich and the poor and people in the rural and urban areas alike.

Ergonomic direction

1. Everyone is not equally tech savvy.
2. Use local language while suing technology.
3. Keep reconfirmation avenue open.

Ergonomics of election 5

OVERVIEW

This chapter gives the readers an overview of the different ergonomic issues associated with election and campaign. The ergonomic issues associated with the entire electoral process are touched upon with directions for improvement from the citizen's perspective. The entire election process, maintenance of peace and cordial relationship among different political parties and the security forces and their ergonomic angles are discussed. The chapter also touches upon the application of ergonomic principles in curbing malpractices in election and how to use ergonomic principles in encouraging citizens to come forward and cast their votes and exercise their rights as citizens.

5.1 ELECTION CAMPAIGN AND VISUAL ERGONOMICS

Election campaign and specially last minute campaigns are vital in attracting citizens in their favour for a political party. People have short-term memory. Thus, any campaign should be routed through visuals supplemented by auditory inputs. For example, posters, banners and wall writings should be in large font size which is easily visible under different illumination level – high and low. Fluorescent paints can be used for areas which are dimly light, and dark backgrounds with light texts are preferable in these areas. All caps should be avoided and where possible upper lowercase should be used. Visuals should be simple, with minimal elements as people would be reading them from moving

DOI: 10.1201/9781003748229-5

FIGURE 5.1 Visual ergonomics in election campaigns

vehicles and would not stand to read them. The placements should factor the height of the people as well as their eye height from different vehicles.

Campaign through road shows should ensure that there should be visuals of the political part in the form of banners and flags only, as leaders would be talking and communicating with the people. It's necessary that leaders mix with the crowd and allow them to talk and vent their problems and even touch them (Figure 5.1).

In political meetings addressing the public, the banners, posters and other visuals should be located all around the venue to designate the exact boundary of the meeting. In digital mediums like televisions and social media, care should be taken to introduce small clips of not more than one-minute duration through which political message could be conveyed. Animation could be used as well for this purpose, preferably in between popular television shows. Extreme care should be taken not to irritate the public as they are keener in watching their favourite shows and thus the duration to be restricted to one minute or less.

Illuminated boards should be brief with minimal text and strategically located so that they do not cause distraction to drivers and thus should not be blinking in highways. On streets and walkways where the public moves, digital displays moving from right to left (if the stereotype is reading from left to right) or from bottom to top (as we scan from top to bottom) is preferable. Ensure there is no flickering else this would cause disturbance to the eyes.

Data of what a party has done how much they have done or not done needs to be conveyed to the public through simple graphics and not statistics. For example, if a party wishes to display how many jobs have been created compared to the previous party instead of numbers they might use two bars – one tall and one short – to show the same. The number of jobs could additionally be written on the bars. The tall one representing the increase in the number of jobs compared to the other party (short bar).

5.2 TOUCHING THE ELECTORATE THROUGH VISUAL ERGONOMICS

If a party is able to touch the hearts of the citizens, the election results tilt in their favour. Convey to the people the developmental works done, how people's life has improved and how society at large has developed. The memory of people is very short lived, and they forget what was before when new developments take place. Politicians should focus on this. For example, through visuals try to show how painful it was for the citizens to travel to office in crowded bus and then through visuals depict how life has become easy for citizens after the introduction of underground railways. When people see two images side by side, they are able to relate and it creates a long-lasting impression in their brains.

Electorates perceive political party as boring and try to avoid them. Engage with the electorate through quizzes, question answer sessions on the party's aim and objective and their election manifesto. Arrange for some small gifts for those who are able to answer correctly. This develops connectivity with people, as they always remember the games they played and the prizes they earned. The small prizes kept at home would help in retrieving the memory of the interaction with the political party and their election manifesto (Figure 5.2).

Visuals should be carefully used not to hurt people's sentiments. For example, in conservative society when poll campaign takes place visuals of women in scanty clothings should be avoided. This irritates people, as they develop a sense that the party has no respect for women. Rather use visuals which touch people. For example, visuals depicting citizens being distributed free winter wear might impress the electorate in a relatively poor area where they can't afford to buy the same. Visuals should refrain from using religion, and it should be customised for people of divergent culture. Here, the political party has to do some homework. For example, the colour black is considered inauspicious in some community. If background research of the political party suggests the same, the same colour should be avoided at all cost when using

FIGURE 5.2 Interacting with the electorate through ergonomics

FIGURE 5.3 Visual ergonomics in touching hearts

visuals like banners, posters and other campaigns in digital medium for these group of people (Figure 5.3).

Citizens are always interested in knowing their leaders in depth and then try to compare themselves with the leader. Thus, through visuals the lifestyle of the leader could be portrayed highlighting how simple, dedicated and honest he is. This type of visual depiction helps citizens to see and synthesise based on their own lifestyle and needs how does the lifestyle of the leader map and whether the leader would be able to make any contribution in developmental activities if voted to power.

5.3 THE ELECTION PROCESS AND THE ERGONOMIC APPROACH

If you look at the election process from a systems or a macro level (discussed in Chapter 1), you would notice certain key elements where ergonomic interventions are necessary. This is because we are dealing with the "people" – our users – who are the most important. When we say user, here the user group comprises the voters, the officials associated directly with the voting process, security personnel in charge of law and order and the people from media who cover the entire voting process.

The voting booths should be easily accessible to varied types of voters. Voters should clearly understand how to vote correctly, and the voting machine should have adequate built-in feedback to indicate whether the voting has taken place and whether the vote has gone to the person the voter wanted to vote. This is where the interface of the voting machine should be minimal and simple. Buttons which are to be pressed should be designed with a bit of concavity so that it mimics users to press with their fingers. The display should clearly indicate what input they have given when they vote for a particular candidate. The list of candidates and their respective symbols should be clear to all. Thus, the user interface should be language independent, area well illuminated and all input and output minimal and clear for elderly, young, literate, illiterate and other variety of users. In cold climate, the button dimension should consider gloved hands and in that case the button surface area should be increased.

The height of the voting booths should be fixed at the elbow height of the voters where they are using the voting machine or stamping on the paper ballot. There should be adequate space around the booth, and for this purpose the elbow-to-elbow distance of the fattest person by spreading (Figure 5.4) the

FIGURE 5.4 Human body dimensions for designing voting booths

upper arm sideways may be taken as a point of reference and a circle drawn around that. In addition, some allowance for one step back, front and sideways should be allowed.

The place where the officials are seated should be properly arranged in the sequence in which any voter has to go through once they enter the voting zone. For example, the first table for identity card verification, the second for cross verification of the voters name in the voting list, the next person for taking the person's signature and so on. This sequence of seating should be maintained just like an assembly line, and it should lead to the voting booth. Once done the voter should exit from another door, thus maintaining a steady flow of voters without any bottleneck.

Labelling on the booth should be at the eye level of the voters while they wait to cast their votes. Any information related to corruption in voting, key person names and phone number should be written near the place where people stand in the queue. If possible the correct way of voting should be visually represented so that voters while standing develop a clear picture of how to vote, and this would prove helpful for elderly voters as it would reconfirm the exact way to vote correctly.

For the voting staffs care should be taken to ensure that the table height is at their elbow height near seating. The chairs should support their back. All staffs should be advised if possible to stand up from the chair, stretch their body for a minute or so and then again take their seat. This would prevent onset of back pain. If possible they can take a stroll for a minute or so.

After the voting is over, many times the ballot boxes or the voting machine needs to be lifted and are very heavy. Here care should be taken to ensure that all heavy objects should have large handles on all the sides so that they could be gripped by all the fingers. While lifting such boxes polling staffs should be advised to bend their knees and keep their back straight to prevent any injury to their back. If there is a need to carry the heavy boxes, it's better to have straps which can then be taken on the shoulder like a rucksack. This distributes the load on the either side of the shoulder and the person can now walk with it.

5.4 CORDIAL RELATIONSHIP IN ELECTION: THE ERGONOMIC ANGLE

It's important that all political parties maintain cordial relationship during the election process so that voters are also at easy in voting. The cordial relationship portrays to the voters an image that voting is going on peacefully and they can participate in it without any fear or hesitation.

First of all, arrangement for seating could be made at strategic locations on the way to polling booths for members of all political parties. Seating should be arranged so that they face each other and facilitate eye-to-eye contact. Arrangement for refreshments should be made to facilitate discussion among the parties. If all parties sit together and discuss with one another, all tension associated with the election process melts down. The very image of different political party members sitting together, talking and eating depicts peaceful voting process and puts the voters at ease. Violence can be curbed to a great extent if such places for interaction among the political parties could be arranged instead of having each party's dedicated places (Figure 5.5).

The ergonomic angle should focus on making the voting process an event to celebrate together with people of different ideology. If necessary movies can also be screened which would entertain the political parties and the voters alike.

Lastly, the people in charge of the election process should clearly depict to all political parties how the voting machine or the ballot boxes are sealed after the voting process and how they are transported and taken to the strong

FIGURE 5.5 Interactive seating arrangements facilitating harmony during elections

room. Thereafter, the entire counting process should be clearly depicted to all political parties.

Once there is adequate transparency in the system people will get an instantaneous feedback for every task performed. Thus, there will be no commotion and confusion as a result of which political harmony will prevail.

5.5 ERGONOMICS IN PREVENTING ELECTION CORRUPTION

Corruptions during elections are an "organised" crime. If this organisation is broken, then the concerted act of the crime diminishes. The human nature is that once transparency is established in the system it becomes difficult to commit the crime because the crime becomes "visible" which criminals do not want. For example, house breaks or thefts happen at night or when everyone is on a holiday. Presence of cameras can deter crimes because in that case the criminals are not only seen but also recorded.

During the voting process as indicated earlier in this chapter, adequate feedback to the voters indicates that their vote has been cast. Similarly when the voting machines or ballot boxes are sealed, the seal should be prominent to all and displayed in a manner to indicate its undercover and protection. The seal should be in such a manner that any attempt to tamper with it would either change the "colour of the seal" (for example) or the form of the seal will change. Both these can be identified from a distance, and thus any attempt to tamper the vote boxes can be identified. Audio and visual feedback could be incorporated in the seals so that any attempt to tamper would illuminate a flickering light along with a loud audio beep which makes detection easy with two layers of redundancy – one visual and the other audio.

During the counting process transparency should be maintained. The counting normally takes place in rounds. After one round of counting the results should be publicly displayed. The final results should be displayed at the end. This builds in trust among the public in system transparency, as they have continuous feedback throughout the counting process.

Lastly, all records related to each voter should be digitally maintained. On demand they should be displayed, and if this is done corruption can be curbed to a great extent. For example, if a party wants to know their booth-wise performance, the same data should be presented to them. This makes them satisfied and they trust the system.

5.6 USING ERGONOMICS TO MAKE AN IMPACT DURING CAMPAIGN

The biggest challenge in a democracy is encouraging people to vote. This responsibility cannot be taken up only by the political party but also by the election commission or similar independent body coordinating the election process.

FIGURE 5.6 Election campaign for encouraging citizens to vote

The campaign should pictorially depict what can happen if people don't vote. They should be explained the power of people and how the country is theirs and it's their responsibility to shape their country. If they don't vote, then it might go into wrong hands and then staying in this country would be extremely difficult. Depict these through visuals and tell people what can go wrong if they don't participate. This gives visual feedback to people and creates in them a mental image of a chaos in the country. Parallelly tell them how they can put things in order by electing a proper leader. Project a rosy picture of the country and how it can be a beautiful place to live in. These creates mental images which makes people act and take decisions accordingly (Figure 5.6).

5.7 KEY POINTS

A. Visual ergonomics of placement of banners, posters and the background, foreground and typography in digital medium augments election campaigns by political parties.
B. Proper visual ergonomics like visuals depicting real-life scenario can touch people emotionally while doing a political campaign.
C. In the election process starting from space layout to workplace design, ergonomics can aid proper crowd flow and effective vote management.
D. Through interactive seating, engaging in entertainment cordial atmosphere can be built up during the voting period.
E. Incorporating adequate feedback, transparency in the system can help fight election relation corruptions.
F. Visual ergonomics of importance of voting and conveying the same through visuals which register fast and better in peoples mind can enhance the participation rate of the people during voting.

5.8 PRACTISE SESSION

A. A political party which has lost the last election is seeking expert advice from an ergonomist as to how to start a campaign which would appeal to the people and give them dividend in the vote box during the election. What would be your advice as a consultant ergonomist?

Ergonomic direction

1. Through visuals project what the ruling party promised and what they could not do.
2. Project to people through audiovisual show/animation what they intend to do if voted to power.
3. Make sure that the campaign should have a couple of keywords with respect to which recall of the party would be easier.
4. The campaign should be repeated in all digital platforms day in and out and also as banners and posters.

B. The election commission of the country "Republic of Telebhaja" seeks ergonomic advise in conducting their first election. The country is just "born" and needs some handholding. What ergonomic advice would you give them?

Ergonomic direction

1. The election venue should factor in people's body dimensions for calculating the furniture dimensions and mapping the space.
2. Information system should be placed at locations where public movement is maximum.
3. Extensive campaign to vote with emotional touch by providing some visuals how the country was born through struggle.
4. Make the voting process transparent and ensure to incorporate feedback when people vote so that they know that they have voted for the right candidate.

C. A country "Khichuri" is finding it tough to eliminate corruption during voting and is unable to curb violence. From an ergonomic perspective what would be your approach for addressing these two issues?

Ergonomic directions

1. Pave the way for a common interactive seating arrangement for all political party.
2. The rate of voting for every booth should be displayed outside.
3. Arrangement for entertainment for all political party with refreshments.
4. Once voting is over the information to be conveyed to all parties and the seals of the voting box should be displayed prominently.

Exercises on application of ergonomics in politics

6

E1. A political party name "Begun Bhaja" is in power for the last 11 years in the country of "Chorchori" which is a small republic. The main opponent of this party is "Kumri" which is trying hard to come to power. The party "Begun Bhaja" now wants to change their strategy for reaching out to the people so that they could remain in power. From an ergonomic view point what should be the course of action of the party "Begun Bhaja"?

Some ergonomic directions

1. List down the developmental works done and gather visuals of the same.
2. Make collage of developmental works done in the last few years.
3. Project some unique work done and a few leaders who could be youth icons.
4. Design a theme song for the political party and start playing it on digital platforms and phone ringtones.
5. Stage street plays and performances on political satires which appeal to the public.
6. As election comes closer the campaign and its repetition should be enhanced.
7. Provide small gifts to the people which are utility product like t-shirts and caps with the logo of the political party engraved. If possible, a timeline of developmental work on the front of the jersey could be printed.

DOI: 10.1201/9781003748229-6

E2. The political party “Dikchao” has been on election campaign before every election. This time they have planned to add some newness in their election campaign. They have been winning a few seats in the parliament but are always short of absolute majority. They would like to apply the principles of ergonomics for connecting with people at all levels of the society. How would they go about?

Some ergonomic directions

1. Make a list of elderly and younger population in the constituency. They are easy to interact with and are emotional.
2. Make campaign posters showcasing emotional elements, for example, “elderly people being treated at home”, “children being taught to love animals” and so on, and how the political party supports the same.
3. Give some clear directions of developmental work with year-wise break up, if voted to power. For example, first year road and infrastructure, second year adult education and so ob. Talk about taking feedback from citizens.
4. Make sure that these campaigns run at least for one year on digital platforms for better impact.
5. If possible, films with story lines, like how a child from a very poor background grows up to be a successful person with the help of the developmental schemes of the government backed by the political party and its ethos, can impact the public mind.

E3. As a consultant ergonomist, you have to analyse the political system in a country “Petebatha”. The country has been under dictatorship and has recently moved into the democratic system of governance. What steps would you follow for conducting an ergonomic analysis?

Some ergonomic directions

1. Identify the elements that comprise the political system.
2. Carefully label where exactly the citizens and politicians stand, the sharp and blunt end.
3. What are the weak links?
4. What is at the core of this system?
5. Variation of users? How to account for
6. Tangible elements
7. Communication aspects

E4. After remaining in power as a ruling party for almost 22 years the political party "Bhalobasha" is facing the election. This time their main competitor is the party "Monerekho". As the election is drawing near, the incidences of political violence are on the mount with significant loss of property and life. The election commission of the country is worried as to how to keep things peaceful. They are now thinking of going for an ergonomic intervention. How should they go about?

Some ergonomic directions

1. Administration should provide equal space to both for campaigning.
2. There should be common forums for discussion.
3. Arrange for friendly meets like football matches with mixed teams.
4. Have neutral campaigns against violence and how it affects people and society and highlight this in public.
5. Regular meeting with law enforcement agency asking everyone to maintain peace along with the citizens. Citizens presence will make the political party behave politely, especially before election!

E5. The party "Oremon" which is in opposition wants to regain power. They are finding it hard to do so as in the last two elections they tried hard but they lost to the ruling party "Orebaba". This is the third time they are fighting the election. From an ergonomic view point can you design a strategy for them wherein their chances of winning the election against the party "Oremon" increase?

Some ergonomic directions

1. List down the touch points where more interaction with people at all levels could be established.
2. Try to identify how, where and what manner the ruling party interacts and communicates with people.
3. Try to see in what manner the communication with people can be better established so as to communicate the parts ethos and agenda.
4. Use multimedia devices in the form of small clips, movies and animation.
5. Make sure the parts ethos and what they can do for people is visible all around and people are exposed to them daily.
6. Refrain from making negative campaign against opposition.
7. Try reaching out to people giving them contact number and names in case of sudden illness, hospitalisation and medical emergencies. It creates positive images.
8. During festival times put banners and promotional materials with high visual content around where maximum movement of people takes place.

E6. The election commission of a country "Bhalobasha" wants citizens to participate in the electoral process and vote for their favourite political party. For the past few years in spite of making all efforts the voter turnout is very low. From an ergonomic view point your help is solicited.

Some ergonomic directions

1. Visually present to citizens how their country is relevant to them and their existence.
2. Through visuals try to present what can go wrong if power goes into wrong hands; it's the people who suffer.
3. Convey to people that they can be safe ad happy only if people around them are honest and good.
4. The basic necessities of life like food, shelter and clothing can be a problem if wrong people control them.
5. Project to them that their children's future is dependent on who rules the country.

E7. There are complaints from some political party that the name and symbol of their political party and candidate is very much at the end of the list at the voting machine and ballot paper. This is leading to fatigue among the citizens and they are ignoring the names at the end of the list and hence voting for those at the beginning of the list. The election commission has asked you to give some ergonomic solutions to this problem.

Some ergonomic directions

1. The allegation raised is not true, because citizens only vote once. Thus, the question of getting fatigue does not arise. The sequence of different candidates and parties in the ballot paper and voting machine normally follows an alphabetical order.
2. When citizens go for voting, they are already decided for which party/candidate they are going to vote. So they are going to look for that name.
3. Even if we accept the allegation of any political party that those at the beginning get more votes, then the only option is to randomise the sequence of the candidates and the party for every ballot paper and in the voting machine. In this way, the biases could be balanced and removed.

E8. The election commission is asking for your advice in deciding upon the dimensions of the voting zone, and the tables and chairs where the staffs work. Your advice is also solicited for facilitating the crowd flow in space.

Some ergonomic directions

1. For tables for the officials working for the election it should be fixed near their elbows while seating.
2. The minimum space of the table surface should be such that the elbows of two persons sitting next to each other should not touch.
3. The cubicle for voting should ensure that the table, on which the voting machine/ballot box and stamping on the ballot paper are kept, should be at the standing elbow height of the citizens. The space around the cubicle should factor the elbow-to-elbow distance while standing. This elbow-to-elbow distance should be calculated by raising the upper arms at the sides and touching the fingers of the palm and then measuring the elbow-to-elbow distance.
4. Height of the cubicle should be in tandem with citizen's stature after wearing shoes.
5. Voters to enter the voting area from one direction and leave from another direction.
6. Control the number of voters entering the voting area, for example, two at a time.

E9. At the time of any election, banners and posters with different visuals play a very important part of any campaign. Political analysts believe that at times these promotional materials do influence some voters to vote in their favour. What are the ergonomic issues that one should keep in mind while designing such campaign materials and while placing them in public domain to ensure they have the maximum effect on the citizens.

Some ergonomic directions

1. Visual cone of people while walking, standing and sitting (e.g., at the bus stand) needs to be considered while pacing the hoardings and banners at correct heights.
2. Consider eye height of people and from that draw a line parallel to the ground. Draw with a protector an angle of 30 degree above this line and 45 degree below this line.
3. Prefer to use upper lowercase for texts as it's easy to read.
4. Use dark text on light background where you feel the illumination level is adequate.
5. Use light text on dark background where the illumination level is always low.
6. The contents of banner should be short and simple so as to be easily readable from moving vehicles.
7. Avoid using pure red and pure green, as those who are colour blind would perceive them in shades of grey.

E10. Some ergonomic directions for handling issues related to politics and election in a country. How should one move ahead? What should be the probable steps?

Some ergonomic directions

1. Systems-level analysis (look for citizen-to-politician communication; modality, how good or bad it is)
2. Citizens are embedded in the system comprising citizen, politicians and bureaucrats and the context.
   Identify the weak link in the above
3. How can the weak link be made robust?
4. Remember two big challenges: every citizen is different; every politician is different
5. How does politicians communicate? Direct, indirect, through work?
6. Game of power – the ruler and opposition
7. The ergonomic aspects of remaining in power and opposition
8. Election: how to encourage people: make them visualise they are part of the society
9. The election process and the context, space and interaction while voting

Bibliography

Acemyan, C. Z., & Kortum, P. (2014, September). To verify or not to verify, that is the question: Voters' use and expectations of voting verification methods. In *Proceedings of the human factors and ergonomics society annual meeting* (Vol. 58, No. 1, pp. 1295–1299). Los Angeles, CA: SAGE Publications.

Austin, J. L. (2019). Towards an international political ergonomics. *European Journal of International Relations*, *25*(4), 979–1006.

Austin, J. L., & Bocco, R. (2016). Becoming a torturer: Towards a global ergonomics of care. *International Review of the Red Cross*, *98*(903), 859–888.

Beazley, M. B. (2013). Ballot design as fail-safe: An ounce of rotation is worth a pound of litigation. *Election Law Journal*, *12*(1), 18–52.

Boudeau, C., Wilkin, P., & Dekker, S. W. (2014). Ergonomics as authoritarian or libertarian: Learning from Colin Ward's politics of design. *The Design Journal*, *17*(1), 91–114.

Bruter, M. (2019). Electoral ergonomics: Three empirical examples of the interface between electoral psychology and design. *Societies*, *9*(4), 82.

Bruter, M., & Harrison, S. (2016). Through the polling booth curtain: A visual experiment on citizens' behaviour inside the polling booth. In A. Blais, J.-F. Laslier, & K. Van der Straeten (Eds.), *Voting experiments* (pp. 323–333). Switzerland: Springer Cham.

Bruter, M., & Harrison, S. (2020). *Inside the mind of a voter: A new approach to electoral psychology*. Princeton, NJ: Princeton University Press.

Budurushi, J. (2016). *Usable security evaluation of EasyVote in the context of complex elections* (Doctoral dissertation, Dissertation, Darmstadt, Technische Universität Darmstadt, 2016).

Byrne, M. D. (2017). Improving voting systems' user-friendliness, reliability, & security. *Behavioral Science & Policy*, *3*(1), 15–24.

Chapanis, A. (1976, July). Ergonomics in a world of new values. In *Proceedings of the human factors society annual meeting* (Vol. 20, No. 1, pp. I–XI). Los Angeles, CA: SAGE Publications.

Dehlinger, J., Harrison, S., & Scala, N. M. (2021). Poll worker security: Assessment and design of usability and performance. In *IIE annual conference. Proceedings* (pp. 698–703). Peachtree Corners, GA: Institute of Industrial and Systems Engineers (IISE).

Ebimgbo, S. O., Nwachukwu, T. S., Ugwu, C. I., Obioji, J. N., Idowu, B. O., & Okoye, U. O. (2022). "They will only remember us during elections": The culture of government support to Nigerian older adults and policy implications. *Journal of Aging & Social Policy*, *36*(5), 767–785.

Elklit, J., & Svensson, P. (1997). The rise of election monitoring: What makes elections free and fair? *Journal of Democracy*, *8*(3), 32–46.

Gable, T. M., Chen, D. W., Darling, C. M., McGlynn, S. A., Kazi, S., Preusse, K. C., … & Schaeffer, L. M. (2016). Heuristic-driven recommendations for improving US voting. *Ergonomics in Design*, *24*(3), 4–8.

Gibson, R. K., Lusoli, W., & Ward, S. (2005). Online participation in the UK: Testing a 'contextualised' model of Internet effects. *The British Journal of Politics and International Relations*, *7*(4), 561–583.

Hsu, J., & Bronson, G. (2018). E-voting technologies usability: A critical element for enabling successful elections. In *Emerging challenges in business, optimization, technology, and industry: Proceedings of the third international conference on business management and technology, Vancouver, BC, Canada 2017* (pp. 61–78). Springer International Publishing AG.

Huckfeldt, R. R., & Sprague, J. (1995). *Citizens, politics and social communication: Information and influence in an election campaign*. Cambridge, MA: Cambridge University Press.

Jackson, N., & Carter, P. (1985). The ergonomics of desire. *Personnel Review*, *14*(3), 20–28.

Jennings, W., & Wlezien, C. (2018). Election polling errors across time and space. *Nature Human Behaviour*, *2*(4), 276–283.

Karp, J. A., & Banducci, S. A. (2001). Absentee voting, mobilization, and participation. *American Politics Research*, *29*(2), 183–195.

Lezaun, J. (2013). Ergonomics of democracy. In K. Cupers (Ed.), *Use matters* (pp. 215–232). London: Routledge.

Moray, N. (2000). Culture, politics and ergonomics. *Ergonomics*, *43*(7), 858–868.

Otte, L. S., & Macht, G. A. (2024, August). Exploring process performance of envelope opening and ballot extraction options in vote-by-mail: A Utah case study. In *Proceedings of the human factors and ergonomics society annual meeting* (p. 10711813241274424). Los Angeles, CA: SAGE Publications.

Pacholski, L., & Piotrowski, K. (2008). Political ergonomics, macroergonomic battles. *Human Factors and Ergonomics in Manufacturing & Service Industries*, *18*(5), 515–524.

Pitcho-Prelorentzos, S., Kaniasty, K., Hamama-Raz, Y., Goodwin, R., Ring, L., Ben-Ezra, M., & Mahat-Shamir, M. (2018). Factors associated with post-election psychological distress: The case of the 2016 US presidential election. *Psychiatry Research*, *266*, 1–4.

Scott, P. A. (2009). Sustainability: An ergonomics watchword for the twenty-first century. In P. A. Scott (Ed.), *Ergonomics in developing regions* (pp. 455–462). Boca Raton, FL: CRC Press.

Sharma, S. (2020). Can't change my political disaffection! The role of political disaffection, trust, and resistance to change in internet voting. *Digital Policy, Regulation and Governance*, *22*(2), 71–91.

Silverstein, M. (2007). Ergonomics and regulatory politics: The Washington State case. *American Journal of Industrial Medicine*, *50*(5), 391–401.

Song, J., & Fiore, S. M. (2020, December). For whom the tale's told: Towards a multidimensional model of targeted narrative persuasion in information operations. In *Proceedings of the human factors and ergonomics society annual meeting* (Vol. 64, No. 1, pp. 1505–1509). Los Angeles, CA: SAGE Publications.

Tannen, R. (2015, September). Intuitive voting: A case study in rapid research & prototype design. In *Proceedings of the human factors and ergonomics society annual meeting* (Vol. 59, No. 1, pp. 1124–1126). Los Angeles, CA: SAGE Publications.

Thatcher, A., Waterson, P., Todd, A., & Moray, N. (2018). State of Science: Ergonomics and global issues. *Ergonomics*, *61*(2), 197–213.

Vicente, K. J. (2008). Human factors engineering that makes a difference: Leveraging a science of societal change. *Theoretical Issues in Ergonomics Science*, *9*(1), 1–24.

Wilkin, P. (2010). The ideology of ergonomics. *Theoretical Issues in Ergonomics Science*, *11*(3), 230–244.

Index

absentee 58
accessories 28
accident 23, 26
accommodate 4, 26, 30, 34
accuracy 35
act 7, 8, 14, 24, 46, 48
action 30, 50
adaptability 34
address 4, 30, 31, 38
administration 2, 3, 5, 9, 14, 23, 52
administrative 2, 3, 5, 8, 24
administrator 5
admissions 26
adult 51
afternoons 22
age 3, 22
aggressive 4
agitated 4
agitations 24
allegation 53
alliance 8
alphabetical 53
ambiguity 2
amenities 5
angle 20, 45, 54
annual 57–59
anthropometry 26
anti-incumbency 18
area 26, 42, 43, 54
arms 54
arrangement 32, 45, 49
arrogant 5
attire 7, 8, 27
audience 30
audiovisual 22, 28, 30, 37, 49
augmented 34
authorities 24

back 5, 10, 13, 16, 29, 44, 45
backed 23, 51
background 31, 34, 38, 42, 48, 51, 54
balanced 53
ballot 43, 45, 46, 53, 54, 57, 58
bank 6, 18, 25, 31
barricades 17
belief 16
belongings 33
bicycle 8
blood 34
bombarded 36
bombs 36
booth 43, 44, 49, 57
bottleneck 2, 8, 44
boxes 45–47
button 43

calamities 26
campaigns 25, 34, 36, 39, 40, 43, 48, 51, 52
candidate 9, 43, 49, 53
checkup 10
citizens 2–11, 13, 16, 20, 21, 23–34, 37–39, 41–43, 47, 51–55, 57, 58
cognitive 14, 23
comfort 5, 8
computer's 9, 20
constituency 9, 51
cooker 23, 24
corruption 1, 9, 23, 25, 28, 44, 46, 47, 49
counting 46, 47
crowded 41

day-to-day 4, 17
decentralised 6
democracy 25, 28, 47, 57, 58
dependent 35, 53
design 34, 38, 48, 50, 52, 57–59
devices 33, 52
dimensions 8, 9, 26, 27, 30, 33, 34, 44, 49, 53
display 9, 41, 43
door 34, 44
drinking 5, 23

ear 31, 36
eating 45

effective 8, 48
elbow-to-elbow 43, 54
electorate 8, 41, 42
elements 1, 13, 25, 26, 39, 43, 51
emergencies 52
ergonomic 1–3, 6, 9, 11–21, 23, 26, 29–34, 37–39, 43, 45, 49–55

fabric 1
facilities 5, 33, 34
family 14, 32
fashion 1, 20
feedback 2, 4, 7–10, 13, 23, 24, 32, 43, 46–49, 51
festival 52
flickering 41, 47
form 17, 23, 24, 31, 35, 40, 46, 52
furniture 9, 26, 32, 49

game 12, 20, 24, 55
gardening 7
gathering 30
gifts 41, 50
God 3
government 2, 9, 11, 17, 23, 34, 37, 51, 57
grassroots-level 32
grievances 4, 23

handholding 8, 49
happening 4, 10
hardships 14
headquarters 13
hierarchy 6, 9, 12, 14
hooliganism 8
human 1, 4, 7–9, 12, 13, 17, 23, 26, 44, 46, 57–59

iconic 8
ideology 17–19, 22, 31, 45, 59
illiterates 33
immunization 34
incidences 23, 26, 52
independent 43, 47
interactive 46, 48, 49

joker 7

knife 34
knowledge 34

labelling 44
ladder 14
leadership 9, 15, 16
locality 2

machine 43, 45, 53, 54
malpractices 39
media 37, 40, 43
meetings 40
minister 12, 13
multisensory 36

Nagasaki 36
newspapers 22
non-availability 35

objective 34, 41
officers 2
oppositions 17, 25
organisation 46
overconfident 16

painful 6, 41
parliament 16, 51
party 2–6, 8–11, 13–26, 29–33, 36–39, 41, 42, 45, 47–53
pyramid 15

questionnaires 31
queue 26, 44

railways 41
refreshment 32
represented 9, 44
resistance 13, 58
rulers 25, 26, 28

Samaritan 14
scan 41
simplicity 8, 18
subdivisions 1
symbolise 7

table 44, 45, 54
tangible 17, 51
teach 18
team 13
threaten 2
transparency 2, 5, 8–10, 13, 23, 46–48

undercover 46
uninterrupted 35

variation 5, 51
ventilation 8
verdict 13
violence 12, 16, 17, 34, 45, 49, 52
voters 10, 43–46, 54, 57

walkways 41
wheelchair 34

year-round 23
yourself 7, 36

zone 44, 53